1 Many creatures hide at the bottom of the sea. Flatfish such as plaice and sole live on the seabed, covering themselves with sand so that they're almost invisible to the sharks and other big fish that hunt them.

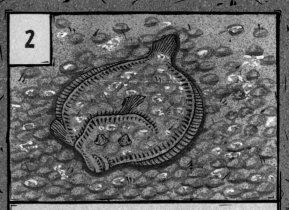

But some flatfish are even better at hiding. Within seconds, flounder can change the colour and the pattern of their skin to look like a particular patch of the seabed — whether it's speckly pebbles or golden sand.

3 Flatfish aren't born on the seabed, though. They hatch out near the surface of the water, and swim here for their first couple of weeks. At this time, they look like most other baby fish.

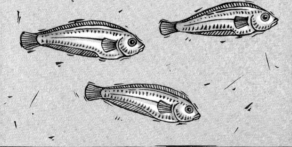

4 Then something extraordinary happens. One of their eyes begins to move round to the other side of their head!

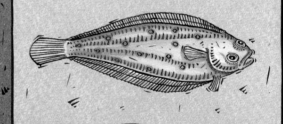

5 After two to three weeks, their mouth has twisted round too, and both eyes are on the top side of their body. The flatfish can't swim upright like other fish now, and they settle on one side on the seabed. Their top side faces upwards, so their two eyes can look up and watch out for danger!

5 If an enemy comes near, the octopus turns pale with fear. It pumps out a cloud of inky liquid to hide itself . . .

3 Octopuses can turn their skin darker or lighter, patterned or plain, bumpy or smooth. The whole thing just takes a second or two.

4 Hiding is a matter of life or death for octopuses. Without their camouflage, they would soon be spotted by razor-toothed enemies such as Moray eels or sharks.

3 OCTOPUSES

6 ...and makes a quick getaway, by sucking water into its body and then squirting it out again — fast!

7 Some kinds of octopus can measure over 4.5 metres from the end of their body to the tip of their tentacles — they would outstretch four children lying head to toe!

8 But this octopus only grows to about 50 centimetres. It's called the Common octopus, and it lives in seas and oceans all around the world.

9 Sometimes, big octopuses will wrap a tentacle around a swimmer. This sounds scary, but they're only being curious. Not all octopuses are harmless, though — Australia's Blue-ringed octopus is deadly poisonous.

CONTENTS

WALKER BOOKS
AND SUBSIDIARIES
LONDON · BOSTON · SYDNEY

DISGUISES AND SURPRISES

Claire Llewellyn

SEABED SURPRISES

1 You'd be lucky to spot an octopus. These strange underwater creatures like to lurk on the seabed or hide themselves on rocks.

2 It's easy for octopuses to hide, because they can change their skin to match their surroundings. Hiding by blending into the background is called camouflage.

SNOW COVER

4 These foxes have to cope with cold snowy weather because they live in lands around the North Pole, at the very top of the world.

1 A beautiful white coat doesn't just help to hide an animal in the snow. This Arctic fox's thick fur also protects it from the cold.

2 Its fur is so thick that heat given off by its body is trapped between the hairs and helps to keep the fox warm.

3 Arctic foxes even have fur on the soles of their feet. This keeps their feet warm and stops them sliding on the slippery ice.

6 ARCTIC FOXES

5 Their white fur helps them to hide from their enemies. Polar bears and wolves both enjoy the taste of Arctic fox.

6 The Arctic fox's camouflage also helps it to hide when hunting its own meals. Its favourite food is hamster-like creatures called lemmings.

1 Polar bears are terrifyingly big and fierce. If a large male bear reared up, it would stand over 3 metres tall. At around 650 kilograms, it weighs more than nine men — and it can run much faster than a man, too!

2 Polar bears live near the North Pole. Their whitish fur keeps them warm and hides them when they hunt other animals — especially seals, their favourite food.

3 Sometimes a bear will lie beside one of the breathing holes that seals make in the ice. If a seal comes up for air, the bear clubs it with its huge hairy paw. Then it grabs the seal in its claws and hauls it out on the ice to eat.

4 Polar bears have few enemies of their own. They are sometimes attacked by Walruses and Killer whales when swimming in the sea, but no wild animals are fierce enough to threaten them on land.

TRICKY TWIGS

1 This isn't a twig, it's a Stick insect! It can stay so still that you could stare at it for hours without spotting it.

2 There are lots of kinds of Stick insect, and they aren't all brown and twiggy like this one. Some are yellowy-gold and look like ripening wheat. Others are green and hide in grass.

3 One kind of Stick insect is the world's longest insect. It's the Indonesian Giant Stick insect and it can grow to be 33 centimetres long.

1 Crocodiles use their camouflage to help them hunt land animals. Their skin looks like bark, and they lie low in the water like heavy logs.

2 But these logs move — and quickly! If something tasty comes by, a crocodile will lunge and grab at it with its jaws. Then it will drag its victim underwater to drown.

4 Stick insects are eaten by all sorts of animals — from birds and lizards, to frogs and snakes. No wonder they have to hide!

5 They aren't hunters themselves, though. Stick insects are plant-eaters and they like to munch on leaves.

6 If they are frightened, some Stick insects play dead and drop to the ground like broken twigs. They crawl away slowly when the coast is clear.

7 But it's not just the insects themselves that are camouflaged. Some kinds of Stick insect lay eggs that look like tiny seeds, and lie hidden on the ground until they hatch.

3 Crocodiles gulp smaller meals down whole, and their jaws are powerful enough to crunch through some animals in one bite. Larger prey are ripped apart and then swallowed bit by bit.

4 The biggest crocodiles are the size of tree trunks! They're the Estuarine crocodiles of South-east Asia and Australia, and they can measure 8 metres or more from snout to tail — as long as two cars!

LEAFY LIARS

1 Although this leafy-looking creature is called an Asian Horned frog, it doesn't really have horns. The pointy bits over its eyes are flaps of skin.

2 From above, the flaps look like leaf tips and add the finishing touch to the frog's disguise. It's almost invisible among the dead leaves on the floor of its rainforest home.

3 The frog has plenty to hide from. Wild cats and owls, and big lizards and snakes are just some of the creatures that eat it for dinner.

4 But unlike most other frogs, Horned frogs have armour. Beneath the skin on their head and back is a bony shield. Many animals find this too tough to bite through.

10 FROGS

5 Asian Horned frogs hide and rest during the day. They hunt for their own food at night, under cover of darkness.

6 At 15 centimetres long, these frogs are big enough to gulp down small lizards and snakes, as well as mice, insects and other frogs. They'll even eat each other!

1 Not all frogs live on the ground — Tree frogs spend almost all their lives up in trees. There are lots of different kinds, and this bright-green one is called a Phyllomedusa.

2 It lives among shiny green leaves in the South American rainforest, and its skin colour helps it to hide from birds and other enemies.

3 Most frogs lay their eggs in water, but not this Tree frog. The female looks for a branch overhanging a pond, and lays her eggs on a leaf.

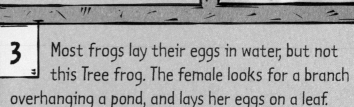

4 She climbs down the tree afterwards, to sit in the pond and let the water soak in through her skin. Then she climbs back up to her eggs — and wees all over them! This makes the eggs swell up, and keeps them damp and jelly-like.

5 In just a few days, the eggs are ready to hatch. The tiny tadpoles drop down into the pond, where they live and grow until they change into frogs — and return to the trees.

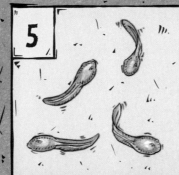

FROGS 11

1 Unlike most spiders, the tiny Crab spider doesn't trap insects in a web. Its colouring helps it to hide in a flower and catch its victims there.

1 Lurking on this flower is a deadly hunter. It's called a Flower mantis and it's only 7 centimetres long. It lives in the warm wet rainforests of South-east Asia.

2 The camouflaged spider lies low and waits. As soon as an insect lands on the flower, the spider grabs and bites it, injecting it with poison so it can't move. Then the hungry spider sucks it dry.

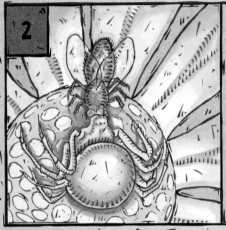

2 This pink beauty is an insect-eating beast. Larger kinds of mantis will even attack small frogs, mice and birds.

3 But flowers don't last forever. When the petals begin to die, the Crab spider has no choice but to pack up and leave. Sometimes it has to move to another plant, where the flowers are a different colour.

4 The spider doesn't want to be seen, of course, so it changes its colour to match the new flower. It's a brilliant trick, but it takes time — somewhere between 5 and 20 days.

12 CRAB SPIDERS

FEROCIOUS FLOWERS

3 If an insect lands nearby, the mantis swivels its head round slowly, keeping its large eyes fixed on its prey.

4 Suddenly, it lashes out. Its front legs snap shut, crushing the insect and holding it fast. There is no escape.

5 Neatly and delicately, the mantis then tears its victim apart. It eats every bit, except the wings and legs — they're just too tough!

6 But mantises are hunted in their turn. Their clever camouflage not only helps them to catch their food, it also hides them from enemies such as lizards and birds.

TREE TEASERS

3 All day long, the gecko hardly moves — not so much as a twitch. But if it is spotted, it tries to frighten its attacker away.

1 You wouldn't want to play hide-and-seek with this gecko. On the branch of a tree, it's so still and so cleverly camouflaged that you'd spend all day trying to find it!

2 It's a Leaf-tailed gecko, and with enemies like snakes and sharp-eyed hawks, it needs its cunning disguise.

The Leaf-tailed gecko lives in the rainforests of Madagascar, an island off the east coast of Africa.

14 GECKOS

4 Although it's only 28 centimetres long, it makes itself look big and scary by puffing itself up, opening its mouth wide — and hissing!

5 But this gecko is a hunter, too. It creeps around after dark, and gobbles up small frogs, as well as moths, beetles and other creepy-crawlies.

1 The Screech owl is another animal that plays hide-and-seek in trees. It lives in the woods and forests of North America, where it keeps itself hidden from dawn to dusk — roosting high up in a tree.

2 Like other owls, the Screech owl hunts at night. Stealthily and silently, it flies through the forest. Other animals rarely hear it coming, because its super-soft feathers muffle the beat of its wings.

3 Very little escapes the owl. Its eyes spot the smallest movement. Its ears catch the faintest sound. Suddenly, it swoops to seize a mouse.

4 All night long, the owl hunts for food. But it returns to its roost as dawn breaks, to spend another day hidden against a tree trunk.

1 Plovers make their nests on open ground — in a gravel pit, perhaps, or on a riverbank or beach. Their speckly eggs are perfectly hidden, because they look just like pebbles.

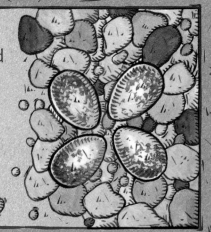

2 The parents take turns to sit on the eggs until they hatch. With enemies like foxes and hawks, it's a dangerous time. The birds may be strongly marked, but when they sit down they seem to vanish into the background.

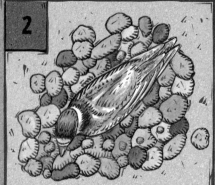

3 The plover chicks' speckly feathers keep them well hidden. But if an enemy comes too close, the plover parents take action.

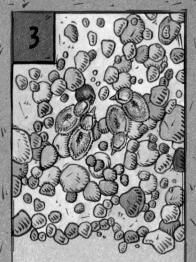

4 One of them runs about, dragging a wing. The plover looks injured, and an easy kill. So the enemy follows, hoping to catch it. Just in time, the plover flies off — having led the enemy well away from its helpless chicks.

16 PLOVERS

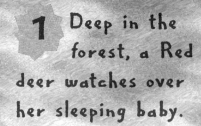

1 Deep in the forest, a Red deer watches over her sleeping baby.

2 The forest is a mixture of sunshine and shadow. The fawn's creamy spots look like patches of sunlight, and they help to hide it in the dappled shade.

3 The fawn's spots will soon fade. And by the time it's two months old, its coat will be just the same colour as its mother's.

BABY CARE

4 Baby animals are easy to kill. The fawn can't run fast enough to escape from danger, so its coat helps it to hide while it grows.

5 Baby deer are born in the thickest part of the forest, where their mothers can keep them safe for the first couple of weeks.

6 Then the mothers lead the fawns back to the herd. Here, the young deer play together, chasing each other among the trees.

7 The games help the deer to grow stronger, and teach them to weave and dodge away from wolves and other hunters.

DEER BARK LIKE DOGS — TRUE OR FALSE?

RED DEER 17

DANGER CHANGER

1 Panther chameleons are crotchety creatures. They like living alone, and they get very angry if another male chameleon invades their patch.

2 They try to scare the intruder away by puffing themselves up, hissing, and turning brilliant shades of yellow and red!

3 Chameleons don't just change colour when they're angry. They go paler if they're too hot, and darker if they're too cold.

1 A hungry chameleon is looking for lunch. Backwards, forwards, up and down — its eyes swivel around. Our eyes work in pairs, but a chameleon's don't. Each eye can move and look in a different direction, and one of them has just spotted a spider!

18 CHAMELEONS

4 Usually, though, Panther chameleons are a blotchy green colour. This is great camouflage, since they spend most of their time up in trees.

5 They live in rainforests on the island of Madagascar, where they're hunted by snakes, as well as eagles and other birds of prey.

6 Chameleons need to hide because they move too slowly to escape from danger. They rarely lift more than one leg at a time, and they make sure they're gripping on tightly before they move the next one.

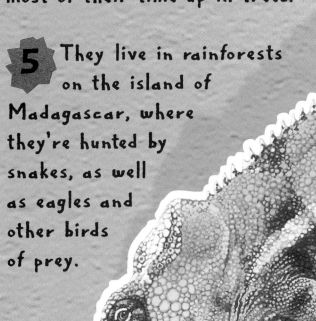

2 Zap! The chameleon shoots out its tongue, hits the spider, and reels it back in. The whole thing is quicker than blinking!

3 The chameleon's amazing tongue is longer than the rest of its body, and it has a sticky, club-shaped tip. The poor spider didn't stand a chance against such a sharp-shooter!

AMBUSH!

1 It's dusk. A hungry lioness crouches in the long dry grass, camouflaged by her sandy-coloured fur.

2 She's well hidden, but too far away to catch her prey. To do that, she needs cunning as well as camouflage.

3 Lions are sprinters — at top speed, they can reach 60 kilometres an hour, but only for 20 seconds or so. Over long distances, an antelope or zebra can easily outrun them.

1 Trapdoor spiders also ambush their prey. They start by making a burrow, using their long, toothed fangs like shovels to dig a hole and carry the dirt away.

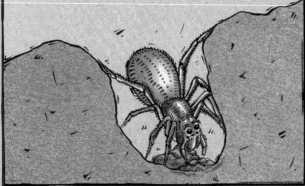

2 After it has lined its burrow with silk, the spider starts work on the door. It spins more silk to glue bits of soil into a lid-shape. Then it spins a hinge so that the lid will open and close.

3 Next the spider crawls outside and scatters bits of grass and twigs over the door, to make it look like the ground around it. The spider's home is hidden now, and its trap is ready!

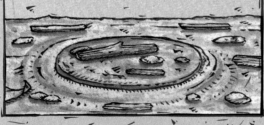

4 The spider crawls back inside its burrow and waits. As soon as it's dark, the spider opens the door a crack, and gets ready to pounce.

5 Sooner or later an insect crawls by. The spider leaps out and bites it, stunning it with poison. Then the spider drags its victim into its burrow. The door slams shut. Now the spider can eat its supper in peace!

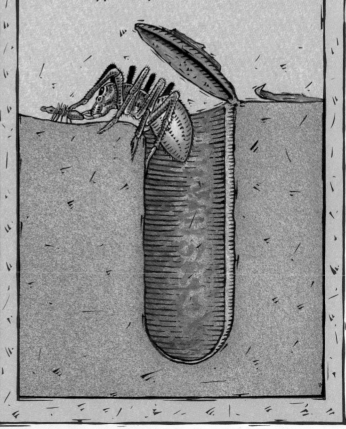

7 Three or four lionesses will sometimes work together to ambush their prey. They growl softly to each other to keep in touch.

8 Lions live in large family groups called prides. There are usually one or two males, five or six females, and a number of youngsters and cubs.

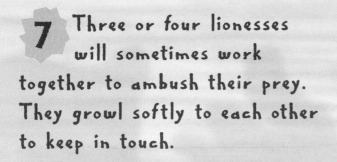

9 Although the lionesses do most of the hunting, it's the male lions that eat first. The cubs are last, and they get such poor scraps that many of them weaken and die.

10 But lions aren't always hunting. They spend most of their time asleep — often as much as 18 hours a day!

4 So the lioness must get nearer before pouncing on her prey. She moves carefully, heading into the wind so that her scent is blown away from the watchful antelope.

6 At last, she charges and springs. Her sharp teeth close round the antelope's throat, choking it to death.

5 The lioness creeps forwards, then freezes... creeps... then freezes... getting ever closer...

INDEX

Consultant: Martin Jenkins

Main illustrations by Priscilla Barrett (20-22); Robin Budden (cover, 10-11); Jim Channell (6-7); Sandra Doyle (12-13); Sarah Fox-Davies (16-17); Robert Morton (18-19); Colin Newman (3-5); Richard Orr (8-9); Clive Pritchard (14-15); picture-strip illustrations by Ian Thompson.

With thanks to Bernard Thornton Artists and the Wildlife Art Agency.

Designed by Beth Aves, Jonathan Hair and Matthew Lilly; edited by Jackie Gaff

First published 1996 by Walker Books Ltd, 87 Vauxhall Walk, London SE11 5HJ

2 4 6 8 10 9 7 5 3 1

Text © 1996 Claire Llewellyn
Illustrations © 1996 Walker Books Ltd

This book has been typeset in Overweight Joe and Kosmic

Printed in Hong Kong

British Library Cataloguing in Publication Data
A catalogue record for this book is available from the British Library.

ISBN 0-7445-2872-0

QUIZ ANSWERS

Page 2 — TRUE
Octopuses have eight tentacles, but their close relatives, squid and cuttlefish, have ten!

Page 6 — FALSE
Some Arctic foxes grow a white coat for winter, and a brown one for when the snow melts in summer. Others live in places where there's little snow and have brown coats all year round.

Page 8 — TRUE
Some kinds of Stick insect have wings and can fly.

Page 10 — FALSE
Some kinds of frog are poisonous and don't need to hide. Their skins are often brightly coloured as a warning.

Page 13 — FALSE
Mantises only live in warm countries.

Page 14 — TRUE
Most geckos have thousands of tiny hooks under their toes, which help them to cling to whatever they touch — handy for walking on the ceiling!

Page 17 — TRUE
Some deer bark to warn of danger.

Page 18 — FALSE
Some kinds of chameleon live on the ground. They're usually a dull brown.

Page 23 — FALSE
Lions grow to 2.5 metres, but tigers can be as long as 2.8 metres!